Process Development Symposium Europe 2019

Frankfurt, Germany
25-26 September 2019

ISBN: 978-1-7138-0693-6

Printed from e-media with permission by:

Curran Associates, Inc.
57 Morehouse Lane
Red Hook, NY 12571

Printed with permission by Curran Associates, Inc. (2020)

For permission requests, please contact AIChE
at the address below.

AIChE
120 Wall Street, FL 23
New York, NY 10005-4020

Phone: (800) 242-4363
Fax: (203) 775-5177

www.aiche.org

Additional copies of this publication are available from:

Curran Associates, Inc.
57 Morehouse Lane
Red Hook, NY 12571 USA
Phone: 845-758-0400
Fax: 845-758-2633
Email: curran@proceedings.com
Web: www.proceedings.com

TABLE OF CONTENTS

KEYNOTE DAY 2: HOW DIGITIZATION CAN ENABLE A CIRCULAR ECONOMY

CREATING A BASELINE FOR LIFELONG LEARNING IN ENGINEERING EDUCATION

PROCESS INTENSIFICATION

CASE STUDIES & LESSONS LEARNED

DIGITIZATION

The Circular Economy Transition: Opportunities and Challenges for Scale-Up and Role Out of New Technology

PROCESS DEVELOPMENT SYMPOSIUM EUROPE

Conference: Process Development Symposium Europe

Year: 2019

Proceeding: 2019 Process Development Symposium Europe

Group: 2019 Process Development Symposium Europe

Session: Keynote Day 1: The Circular Economy Transition

Time: Monday, May 27, 2019 - 8:15am-9:00am

Authors:

Waas, M., Nouryon

Over the last decade new/adopted sustainable technologies have been developed based on biochemistry, electrochemistry and digitization. At the moment these solutions are at lab, Pilot or Demo scale. More sustainable energy and heat have become available and this trend will continue. The Paris agreement and subsequent national responses have increased the opportunities and the need for sustainable technologies. The challenge for Circular Economy transition now lies in scaling up the technologies and develop new business models, value chains, partnerships, and financing methods.

Bringing Data Analytics and ML to the Shop Floor for Better Process Performance

Conference: Process Development Symposium Europe

Year: 2019

Proceeding: 2019 Process Development Symposium Europe

Group: 2019 Process Development Symposium Europe

Session: Process Optimization

Time: Monday, May 27, 2019 - 10:00am-10:30am

Authors:

Cura, M., Optimistik

Machine Learning tools and algorithms are continuously evolving and provide new opportunities to support and augment the capabilities of staff in the plant. Even though, several challenges need to be overcome to achieve that goal:

- Continuous Data collection from multiple and heterogenous sources to provide the required data streams.

- Data structuration and combination to provide meaningful and reliable process information.

- Integrate business expertise based on process modeling to enrich the information available.

- Provide operational teams and process experts with tools, leveraging on state of the arts Machine Learning algorithms, properly suited to answer specific and recurring time-consuming business cases and usages.

We will go through the different steps required to succeed in this kind of project and share the experience acquired in various process industries (chemicals, food and beverage, metallurgy...).

Although technical solutions can support this process, a proper approach focusing on teams needs and usages, autonomy and agility is required to bring a sustainable and operational solution.

Decision support in Industrial Applications: Multi-Criteria Optimization at its Best

PROCESS DEVELOPMENT SYMPOSIUM EUROPE

Conference: Process Development Symposium Europe

Year: 2019

Proceeding: 2019 Process Development Symposium Europe

Group: 2019 Process Development Symposium Europe

Session: Process Optimization

Time: Monday, May 27, 2019 - 9:30am-10:00am

Authors:

Bortz, M., Fraunhofer Institute for Industrial Mathematics (ITWM)

Process design is a multicriteria optimization problem: Costs as small as possible, quality indicators as high as possible shall be achieved. This requires calculating the Pareto boundary, which makes many model evaluations necessary and thus is a computationally demanding task. Once the Pareto boundary has been determined, it can be used for an interactive decision support, which enables the engineer to balance the objectives e.g. increase quality at competitive total costs.

In this contribution, we review computationally highly efficient methods to resolve the Pareto boundary within a predefined accuracy. Among these are adaptive scalarization (Bortz et al, 2014) and exploration schemes (Heese et al, 2019), where the latter employ machine-learning methods. Furthermore, the application of multicriteria approaches in the presence of parametric model uncertainties is described (Bortz et al, 2017).

The benefit of applying multicriteria optimization in process design is reported for different industrial applications. The transferability of the method to data reconciliation and the detection of faulty measurements is demonstrated as well.

M. Bortz, J. Burger, N. Asprion, S. Blagov, R. Böttcher, U. Nowak, A. Scheithauer, R. Welke, K.-H. Küfer, and H. Hasse, 2014, Comp. Chem. Eng., 60, 354–363

M. Bortz, J. Burger, E. von Harbou, M. Klein, J. Schwientek, N. Asprion, R. Böttcher, K.-H. Küfer, and H. Hasse. Industrial & Engineering Chemistry Research, 56(44):12672-12681, 2017

R. Heese, M. Walczak, T. Seidel, N. Asprion, M. Bortz, 2019, Comp. Chem. Eng. 124, 326-342

Machine Learning meets Process Simulator: Grey Box and Surrogate Modeling to Support Process Optimization

Conference: Process Development Symposium Europe

Year: 2019

Proceeding: 2019 Process Development Symposium Europe

Group: 2019 Process Development Symposium Europe

Session: Process Optimization

Time: Monday, May 27, 2019 - 9:00am-9:30am

Authors:

Asprion, N., BASF SE

Besides experimentation process modeling, simulation and optimization is one of the key elements of process development. Especially, when comparing process concepts with regard to partly competing sustainability criteria, multicriteria optimization and decision support (cf. for example Bortz et al., 2014, Burger et al., 2014) are indispensable. The applicability of such methods is sometimes restricted either due to missing models for certain process steps to complete a simulation model for the whole process or by costly and hardly solvable simulation models. In both cases machine learning can provide a feasible solution to tackle these restrictions.

For missing scientific process models available process data can be used to set-up data-driven models. In most cases process data are not or only partly available at the interfaces of the process step to be modeled. Therefore, as a first step usually a simplified model for the process step is included in the simulation model to complete the model for the whole process and used as a soft-sensor for the missing information. After this step the necessary information is available to model the process step for example with machine learning methods separately (so-called incremental model identification, cf. e.g. Bardow and Marquardt, 2004) and include it afterwards in the entire process simulation model.

In cases where a process model is too hard to be applied directly for optimization, the use of surrogate modeling (Bhosekar and Ierapetritou, 2018) can be a resort. Here the original model is sampled, and the collected data of these samples can be taken to train for example artificial neural networks reflecting the optimization variables as inputs and the objectives and constraints as outputs. The obtained optimized variables can be used as inputs for the original process model to verify the results of the surrogate model. If the agreement with the predicted results of the surrogate is not satisfying the verification results can be used to retrain the surrogate, run again the optimization and repeat the verification step. This iterative procedure could be repeated until sufficient agreement is obtained.

Both methods will be demonstrated for a process simulation example of a Cumene process.

A. Bardow, W. Marquardt, 2004, Chem. Eng. Sci., 59, 2673 – 2684.

M. Bortz, J. Burger, N. Asprion, S. Blagov, R. Böttcher, U. Nowak, A. Scheithauer, R. Welke, K.-H. Küfer, and H. Hasse, 2014, Comp. Chem. Eng., 60, 354–363. doi:10.1016/j.compchemeng.2013.09.015

J. Burger, N. Asprion, S. Blagov, R. Böttcher, U. Nowak, M. Bortz, R. Welke, K.-H. Küfer, and H. Hasse, 2014, Chem. Ing. Tech., 86 (7), 1065–1072. DOI: 10.1002/cite.201400008

A. Bhosekar, M. Ierapetritou, 2018,Comp. Chem. Eng. 108, 250 – 267.

Agrochemical Process Research: Searching For The Holistic Solution

PROCESS DEVELOPMENT SYMPOSIUM EUROPE

Conference: Process Development Symposium Europe

Year: 2019

Proceeding: 2019 Process Development Symposium Europe

Group: 2019 Process Development Symposium Europe

Session: Process Optimization Continued

Time: Monday, May 27, 2019 - 11:30am-12:00pm

Authors:

Brohm, D., Bayer AG

The substitution of current feedstock with renewable resources is a major challenge in today's chemical process industries. To focus more specifically on this research area, the Collaborative Research Center SFB/TR63 – Integrated Chemical Processes in Liquid Multiphase Systems (InPROMPT) of the Deutsche Forschungsgemeinschaft (DFG) was conceived. The main emphasis of this collaboration is the development of process structures for homogeneously catalyzed reactions in liquid multiphase systems.

As part of this effort, we primarily focus on the development of optimal reactor concepts for liquid multiphase systems, in particular for the rhodium-catalyzed hydroformylation of long-chain olefins. The main challenge of this process is the quantitative recovery of the rhodium catalyst and equally important ligands. To this end, thermomorphic solvent systems (TMS) are used which allow for temperature induced phase splitting in the post-reaction mixture [1]. The catalyst rich phase is recycled to the reactor and the product rich phase is treated further downstream. Applying a multi-scale design approach, the molecular design of the solvent system directly influences all other process units. This leads to an overall process design task that combines both the solvent and reactor optimizations as part of an integrated process optimization problem.

This is no simple task and this presentation focuses on our recent achievements and methods used towards reaching this goal. As an expansion to the single-stage separation TMS, an extraction cascade was proposed to further reduce catalyst leaching and increase the domain of acceptable TMS solvents [2]. This prompted the development of several strategies for TMS solvent design [3] with an emphasis on not only process performance but also considering ecological characteristics [4]. Frequent liquid-liquid equilibrium (LLE) calculations during optimization are numerically challenging and in order to simplify the optimization of the resulting process flowsheets, several different surrogate models representing the LLE and catalyst partitioning were implemented [2,5]. On the reaction side, The use of the EPF methodology [6] allows us to identify reactor concepts, such as a cyclic semi-batch or helically coiled tubular reactor, that lead to better reaction performance than when using conventional arrangements [7,8]. The various tools and strategies being devised in this project will hopefully enable us to develop new processes based around complex homogeneously catalyzed multiphase reactions, such as for the hydroformylation of long-chain olefins, as part of a single, integrated design step.

Analysis and Optimization of Complex Flowsheet Simulations with Parallel Computing and Machine Learning

Conference: Process Development Symposium Europe

Year: 2019

Proceeding: 2019 Process Development Symposium Europe

Group: 2019 Process Development Symposium Europe

Session: Process Optimization Continued

Time: Monday, May 27, 2019 - 12:00pm-12:30pm

Authors:

Schöneberger, J., Chemstations Europe GmbH
Fricke, A., Chemstations Europe GmbH

Convergence problems are common when dealing with complex flowsheet simulations. These usually do not result from shortcomings of the numerical algorithms, but from an unfavorable or even impossible setting of the degrees of freedom for the underlying system of equations.

Due to the highly non-linear relationships of the variables of state, users of simulation software often cannot identify which design variables in which range or combination lead to unsolvable systems of equations. This leads to time-intensive trial-and-error simulation runs in practice.

Multivariate sensitivity studies (MSS) can be used to systematically and automatically analyze complex flowsheets in the multidimensional space of the design variables. Parallelization of the calculation of flowsheets drastically reduces the time required.

Machine learning methods can be applied, for instance, to identify and exclude areas where the flow diagram does not converge before the calculation. This artificial intelligence is used to control optimization algorithms for the evaluation of complex flow diagram simulations.

This approach is demonstrated using two examples from the field of distillation. Figure 1 shows the results space of an MSS for a rectification column for the separation of ethanol and water.

Process Design and Optimization for the Hydroformylation of Long-chain Alkenes

Conference: Process Development Symposium Europe

Year: 2019

Proceeding: 2019 Process Development Symposium Europe

Group: 2019 Process Development Symposium Europe

Session: Process Optimization Continued

Time: Monday, May 27, 2019 - 11:00am-11:30am

Authors:

McBride, K., Max Planck Institute for Dynamics of Complex Technical Systems

The substitution of current feedstock with renewable resources is a major challenge in today's chemical process industries. To focus more specifically on this research area, the Collaborative Research Center SFB/TR63 – Integrated Chemical Processes in Liquid Multiphase Systems (InPROMPT) of the Deutsche Forschungsgemeinschaft (DFG) was conceived. The main emphasis of this collaboration is the development of process structures for homogeneously catalyzed reactions in liquid multiphase systems.

As part of this effort, we primarily focus on the development of optimal reactor concepts for liquid multiphase systems, in particular for the rhodium-catalyzed hydroformylation of long-chain olefins. The main challenge of this process is the quantitative recovery of the rhodium catalyst and equally important ligands. To this end, thermomorphic solvent systems (TMS) are used which allow for temperature induced phase splitting in the post-reaction mixture [1]. The catalyst rich phase is recycled to the reactor and the product rich phase is treated further downstream. Applying a multi-scale design approach, the molecular design of the solvent system directly influences all other process units. This leads to an overall process design task that combines both the solvent and reactor optimizations as part of an integrated process optimization problem.

This is no simple task and this presentation focuses on our recent achievements and methods used towards reaching this goal. As an expansion to the single-stage separation TMS, an extraction cascade was proposed to further reduce catalyst leaching and increase the domain of acceptable TMS solvents [2]. This prompted the development of several strategies for TMS solvent design [3] with an emphasis on not only process performance but also considering ecological characteristics [4]. Frequent liquid-liquid equilibrium (LLE) calculations during optimization are numerically challenging and in order to simplify the optimization of the resulting process flowsheets, several different surrogate models representing the LLE and catalyst partitioning were implemented [2,5]. On the reaction side, The use of the EPF methodology [6] allows us to identify reactor concepts, such as a cyclic semi-batch or helically coiled tubular reactor, that lead to better reaction performance than when using conventional arrangements [7,8]. The various tools and strategies being devised in this project will hopefully enable us to develop new processes based around complex homogeneously catalyzed multiphase reactions, such as for the hydroformylation of long-chain olefins, as part of a single, integrated design step.

Black Box Modelling - The Fastest Way to Optimize Processes

PROCESS DEVELOPMENT SYMPOSIUM EUROPE

Conference: Process Development Symposium Europe

Year: 2019

Proceeding: 2019 Process Development Symposium Europe

Group: 2019 Process Development Symposium Europe

Session: Modeling & Data Analytics in Process Optimization

Time: Monday, May 27, 2019 - 2:30pm-3:00pm

Authors:

Krüger, A., Atlan-Tec Systems Gmbh

For over 25 years we have had good experience with black box models, in our case special hybrids of artificial neural networks (ANN) and CG algorithms. The approach provides us with a very effective method to model the actual behavior of a process. Unrealistic assumptions and the search for an explanation why the theory is not confirmed in practice in this particular case are omitted. And above all, it is possible to investigate connections for which there is no usable theoretical approach yet.

Here are a few examples:

- When it came to the basis weight of recycled paper, our model was the fastest way to create a soft sensor.

- In multi-stage polymerization with extrusion, we predicted material properties such as notched impact strength with very high accuracy and were able to optimize the formulation from the first polymerization stage to the additives on the extruder in order to find the most cost-effective way to achieve the desired impact strength. Without the data-driven analysis, it would have been necessary to first understand the relationship between molecule structure and notched bar impact strength and then the relationship between formulation, process parameters and molecule structure.

- For a similar process, there was a viscosity soft sensor for the polymer based on a fairly good theoretical approach. After several years, however, he did not deliver correct values and no expert could be found who knew the model well enough and could adapt the calculation. The soft sensor was then replaced by a purely empirical model, which is constantly updated automatically with current data.

- During a combustion process, we were able to produce good NOx models even though the fuel consisted of a mixture of substances that was difficult to describe and unknown amounts of air flowed into the combustion chamber. We have made it possible to improve the efficiency and durability of the boiler through optimal air supply.

The producing Industry creates new valuable Products from raw materials, beside it generates a lot of Data. These Data combined with Process Knowledge and Laboratory Data give us the possibility to improve our process efficient to new limits.

The Grey-Box Approach provides us with a very effective and efficient method to improve a process or a process step.

Optimization of Batch and Continuous Processes with Grey Box Modelling

Conference: Process Development Symposium Europe

Year: 2019

Proceeding: 2019 Process Development Symposium Europe

Group: 2019 Process Development Symposium Europe

Session: Modeling & Data Analytics in Process Optimization

Time: Monday, May 27, 2019 - 2:00pm-2:30pm

Authors:

Kulinna, H., BASF Lampertheim Gmbh

The producing Industry creates new valuable Products from raw materials, beside it generates a lot of Data. These Data combined with Process Knowledge and Laboratory Data give us the possibility to improve our process efficient to new limits.

The Role of Process Engineering in the Digital Transformation

Conference: Process Development Symposium Europe

Year: 2019

Proceeding: 2019 Process Development Symposium Europe

Group: 2019 Process Development Symposium Europe

Session: Modeling & Data Analytics in Process Optimization

Time: Monday, May 27, 2019 - 1:30pm-2:00pm

Authors:

Jung , N., AVEVA

While Digital Transformation hits every corner of industry, the process engineering discipline has been largely excluded from this trend so far. PwC foresee that for process licensors / EPCs, data driven digital services can displace process technology as a differentiator1).

The objective of this presentation is to provide an overview of the obstacles to Digital Transformation for the process discipline and explain how these can be overcome. A special emphasis will be given on the role of the process simulation tool (white box modelling) as a catalyst for transformational change.

The presentation will examine several challenges specific to process engineering:

1) According to DECHEMA, one major obstacle for digital transformation is the division of process simulators into single-purpose point solutions. Today, separate models may be created – often in different tools – for process design, control strategy design, operator training simulation, performance monitoring and online optimization. This drives up the total cost of modelling to a degree than can become prohibitively expensive. One single model should be able to cover the entire plant lifecycle from idea to operations2.

2) Engineering disciplines, other than process, collaborate closely with database-driven tools and assured workflows. Process simulators are typically poorly integrated into this workflow, and if so, with a single directional information flow. This lack of integration makes it nearly impossible to break down silos and enable new ways of working, like the agile development methods that have transformed software engineering.

3) Over the decades, legacy process simulators got overloaded with niche features and functions only usable by experts. This bloat makes them not suitable for automation and operations.

4) The potential benefits of Artificial Intelligence and Machine Learning for process engineering are not widely understood. Appropriate black box tools are not deeply embedded within white box tools.

More recently, stakeholders see the Digital Twin as the most important building block for Digital Transformation of the process industries. The process discipline must be a central part to this transformation representing the process behavior of the Digital Twin3).

While legacy simulators are well-suited to accurately simulate processes, their decades-old architectures mean they are not ideal to serve the entire plant lifecycle and support the Digital Transformation. We will use the AVEVA SimCentral Simulation Platform4) as an example of how the identified obstacles can be overcome with a next generation process simulator. Case examples from leading EPC and Operating companies will be outlined and will serve as a way to explain the related benefits.

1) Digital business models in plant engineering and construction in an international comparison, A benchmarking study of PwC and VDMA, May 2019 (link)

2) DECHEMA Whitepaper Digitalisierung in der Chemieindustrie, September 2016 (link)

3) DECHEMA Tutzingen Thesen 2018 (link in english)

4) AVEVA SimCentral Simulation Platform (link)

Risk-based Steering of New Process Technology Development

Conference: Process Development Symposium Europe

Year: 2019

Proceeding: 2019 Process Development Symposium Europe

Group: 2019 Process Development Symposium Europe

Session: Risk Mitigation in Process Development

Time: Monday, May 27, 2019 - 3:30pm-4:00pm

Authors:

Huizenga, P., Shell Global Solutions International

For over 25 years we have had good experience with black box models, in our case special hybrids of artificial neural networks (ANN) and CG algorithms. The approach provides us with a very effective method to model the actual behavior of a process. Unrealistic assumptions and the search for an explanation why the theory is not confirmed in practice in this particular case are omitted. And above all, it is possible to investigate connections for which there is no usable theoretical approach yet.

Here are a few examples:

- When it came to the basis weight of recycled paper, our model was the fastest way to create a soft sensor.

- In multi-stage polymerization with extrusion, we predicted material properties such as notched impact strength with very high accuracy and were able to optimize the formulation from the first polymerization stage to the additives on the extruder in order to find the most cost-effective way to achieve the desired impact strength. Without the data-driven analysis, it would have been necessary to first understand the relationship between molecule structure and notched bar impact strength and then the relationship between formulation, process parameters and molecule structure.

- For a similar process, there was a viscosity soft sensor for the polymer based on a fairly good theoretical approach. After several years, however, he did not deliver correct values and no expert could be found who knew the model well enough and could adapt the calculation. The soft sensor was then replaced by a purely empirical model, which is constantly updated automatically with current data.

- During a combustion process, we were able to produce good NOx models even though the fuel consisted of a mixture of substances that was difficult to describe and unknown amounts of air flowed into the combustion chamber. We have made it possible to improve the efficiency and durability of the boiler through optimal air supply.

The producing Industry creates new valuable Products from raw materials, beside it generates a lot of Data. These Data combined with Process Knowledge and Laboratory Data give us the possibility to improve our process efficient to new limits.

The Grey-Box Approach provides us with a very effective and efficient method to improve a process or a process step.

Shell's Gas-to-Liquids (GTL) Technology and Opportunities in Energy Transition

Conference: Process Development Symposium Europe

Year: 2019

Proceeding: 2019 Process Development Symposium Europe

Group: 2019 Process Development Symposium Europe

Session: Risk Mitigation in Process Development

Time: Monday, May 27, 2019 - 4:30pm-5:00pm

Authors:

van Bavel, S., Shell

Shell's gas-to-liquids (GTL) technology converts natural gas – the cleanest-burning fossil fuel – into high-quality liquid products that would otherwise be made from crude oil. These products include transport fuels, motor oils, process oils and the ingredients for everyday necessities like detergents and cosmetics. GTL products are colorless, odorless and contain almost none of the impurities (sulphur, aromatics and nitrogen) that are found in crude oil.

Shell has deployed its GTL technology in two world-scale commercial plants in Malaysia (Shell Middle Distillate Synthesis) and in Qatar (Pearl GTL, which is the world's largest GTL plant). Shell has invested more than $1 billion and filed over 3,500 patents in developing the gas-to-liquids process and continues with developments to make GTL technology more affordable, more competitive and more energy efficient.

GTL technology could be an enabler for making low-carbon products, such as low-carbon aviation fuel, low-carbon diesel for heavy-duty applications, low-carbon chemicals and specialties, by weaving into the GTL process (intermittent) green hydrogen and CO_2 as feedstock.

In such "Hybrid GTL" concepts, (intermittent) green hydrogen is utilized during the "day" when renewable power is available, while natural gas is utilized during the "night", like in classical GTL. This eliminates a need for costly large-scale storage (in form of battery or hydrogen storage), which is required for PTL (power-to-liquids). Natural gas becomes a way of addressing intermittency of renewables in Hybrid GTL. Compared to PTL, the green element of a Hybrid GTL plant is much more flexible, making such concepts more affordable and fitting in well with intermittent renewable power or intermittent green hydrogen.

Sources of CO_2 could include CO_2-rich off-gases from internal recycles, external point sources and air capture. Building blocks for converting CO_2 to CO include reverse water gas shift, CO_2 electrolysis, dry reforming and methanation followed by partial oxidation of methane.

As a transition path from today's solutions based on natural gas to fully decarbonized (PTL) solutions on a longer term, Hybrid GTL is one of the elements in the mosaic of solutions in Energy Transition with a potential for early deployment.

Use of a Mixed Solvent Thermodynamic Framework to Predict the Formation of Amine-Hydrochloride Salts in Crude Tower Overhead Systems

Conference: Process Development Symposium Europe

Year: 2019

Proceeding: 2019 Process Development Symposium Europe

Group: 2019 Process Development Symposium Europe

Session: Risk Mitigation in Process Development

Time: Monday, May 27, 2019 - 4:00pm-4:30pm

Authors:

Gerbino, AJ, OLI Systems

Ionic modeling for refinery overhead unit operations with rigorous electrolyte thermodynamics.

OLI System's Mixed Solvent Electrolyte Model (MSE) has been used to develop a framework to obtain a comprehensive understanding of physical properties, thermochemical behavior and phase equilibria of amines and their hydrochloride salts. This presentation will discuss the capabilities of this framework in an industry-standard process simulator to simulate the contributions of ionic and non-ionic species in refinery overhead units. The MSE model will be utilized to successfully identify water dew point, salt point and an ionic dew point for systems involving neutralizing amines. This simulation technology offers the possibility of understanding and eliminating the phenomena that lead to corrosion in refinery overheads and estimating optimized operating conditions for the system.

How Digitalization Can Enable a Circular Economy

PROCESS DEVELOPMENT SYMPOSIUM EUROPE

Conference: Process Development Symposium Europe

Year: 2019

Proceeding: 2019 Process Development Symposium Europe

Group: 2019 Process Development Symposium Europe

Session: Keynote Day 2: How Digitization Can Enable a Circular Economy

Time: Tuesday, May 28, 2019 - 8:15am-9:00am

Authors:

Falter, W., Deloitte Consulting

Unlike discrete manufacturers, process industries are confronted with specific issues to optimize the utilization of their assets and invested capital in order to maximize profit per hour rather than profit per unit of product. Market, technical, regulatory/ compliance, reliability/ integrity and other issues keep those manufacturers from profit maximizing their processes.

Digitalization and especially Real Time Optimization help those companies to take their performance to the next level. Several tools and approaches are available to achieve those targets.

In spite of those obvious benefits to increase profit per hour, adoption of those digital practices has been disappointingly slow and small so far. Data transparency and dis-incentives have been the major roadblocks to implement digitalization in process industries faster and more successfully. Paying attention to those issues can facilitate and accelerate the adoption of digitalization in process industries.

Closing the Gap between Process Engineering and Safety with a Safety-Focused Master's Degree

Conference: Process Development Symposium Europe

Year: 2019

Proceeding: 2019 Process Development Symposium Europe

Group: 2019 Process Development Symposium Europe

Session: Creating a Baseline for Lifelong Learning in Engineering Education

Time: Tuesday, May 28, 2019 - 10:00am-10:30am

Authors:

Hecht, K. J., University of Utah
Krause, U., Otto-von-Guericke-Universität

In the industrial practice, safety is a top priority for process engineers however, this is not typically the case in education programs for chemical and process engineers. At some universities process engineers graduate having heard little or even nothing about process safety. Consequently, in their professional lives they have two options to tackle process safety issues: training on-the-job or leaving process safety issues entirely to specialists or external consultants.

Health, safety, and environmental (HSE) specialists manage a range of topics including occupational safety and health, hazardous materials management, environmental protection, and legal compliance. Hence, HSE managers have to be all-rounders with a solid engineering background, but inevitably they will have limited knowledge concerning the work of the process engineers.

To bridge the gap between the process engineer and HSE manager, the Otto-von-Guericke University in Magdeburg, Germany introduced the Process Safety and Environmental Engineering (PSEE) master program in 2016. Students from around the world with a bachelor degree in chemical or process engineering receive training in the natural scientific fundamentals of technical processes, especially those related to safety and environment so that they can think and act holistically in the assessment of safety and environmental concerns and their prevention and mitigation.

This contribution discusses the role of safety in process development and if there is an "optimal" level of safety training for process engineers. The PSEE curricula is compared to other concepts for educating process engineers in safety and examples of teaching activities that incorporate safety analysis into process engineering activities are shown.

Integrating Global Simulation Competition into Senior Design

PROCESS DEVELOPMENT SYMPOSIUM EUROPE

Conference: Process Development Symposium Europe

Year: 2019

Proceeding: 2019 Process Development Symposium Europe

Group: 2019 Process Development Symposium Europe

Session: Creating a Baseline for Lifelong Learning in Engineering Education

Time: Tuesday, May 28, 2019 - 9:00am-9:30am

Authors:

Halpern, J. M., University of New Hampshire

Process Simulation Cup (PSC) is an international process simulation competition associated with Chemistations ChemCAD software. As part of this competition, a control system needs to be optimized for an already designed process. Typically, the simulation system, can be optimized through trial and error without regard of design of experiments, safety, or real-world implications. As part of the learning objectives in Senior Process Design, the students were asked to engage the limitations and implications of their answers. As such, safety statements were expanded on beyond the answer given in the international competition. The final requested written response engaged an integration of process control, safety, economics, and process design.

A friendly, game oriented, competition got the students involved and excited. Yet, many students failed the learning objectives, and failed to identify implications of their answer in a safety paradigm. In other words, many students pushed the limits of the process design, yet minor fluctuations, as one would expect on a process line, would overload the system. Few students reported these potential fluctuations, and therefore, did not understand the full repercussions of their answer.

Reported in this presentation will be typical student answers, grading sheets, learning objectives, and how the problem has been integrated into my course.

Short Bio: Prof. Halpern earned his PhD in Chemical Engineering in 2010 at Case Western Reserve University in Chemical Engineering. His first postdoc (2011-2012) was funded through an NIH NRSA fellowship at Case Western Reserve University in Biomedical Engineering. Prof. Halpern's second postdoc was funded through Fulbright and Lady Davis Fellowships (2013-2014) to work in Israel at the Technion in the Department of Chemical Engineering. He joined the Department of Chemical Engineering at the University of New Hampshire in 2014 and currently heads the SEEDS (Surface Enhanced Electrochemical Diagnostic Sensors) laboratory.

Playful Exploration and Intuitive Understanding of Chemical Processes by Using Process Simulation

Conference: Process Development Symposium Europe

Year: 2019

Proceeding: 2019 Process Development Symposium Europe

Group: 2019 Process Development Symposium Europe

Session: Creating a Baseline for Lifelong Learning in Engineering Education

Time: Tuesday, May 28, 2019 - 9:30am-10:00am

Authors:

Rarey, J., DDBST GmbH

Since 2012, a special MSc. Course "Chemical Process Development" has been offered to Chemistry students and foreign guests at the University of Oldenburg. Course duration is 8 weeks full day. Instead of homework, students are solving many different workshop problems in the class. The lecturers are present and available for help and discussions at any time. Group work, discussions and creative extension of the tasks are encouraged.

During the first 4 weeks, students are introduced to simulator software (both CHEMCAD and Aspen Plus), chemical thermodynamics with special focus on the models and methods in the thermodynamic engines of the simulator software, conceptual design of separation processes and pinch technology. To better understand the working of the software, one week is spend on lectures and workshops to teach practical skills in scientific programming and numerical methods. The second half of the course features novel design projects, which are worked by groups of two (sometimes three) students.

The performance of the students is judged from the observations during the course, a seminar presentation and a written report. The lectures conduct intensive presentation training with special focus on project pitching in a commercial environment.

The course is taught in the English language and hosted participants from South Africa, Thailand and Oman over the years. Several participants extended their projects within internships of 3-month duration in these countries.

The format results in a high motivation of the participants. Learning and working in a team are strongly encouraged.

Results and observations from the course will be presented in the presentation.

Modeling and Operation of Enzymatic Catalyzed Reactive Wall Columns

Conference: Process Development Symposium Europe

Year: 2019

Proceeding: 2019 Process Development Symposium Europe

Group: 2019 Process Development Symposium Europe

Session: Process Intensification

Time: Tuesday, May 28, 2019 - 11:30am-12:00pm

Authors:

Egger, T., Hamburg University of Technology

The enzymatic catalyzed reactive dividing wall column (eRDWC) is a highly integrated apparatus that combines the advantages of enzymatic catalyst, reactive distillation and dividing wall column technology. Compared to a sequence of lower integrated apparatuses, an eRDWC thus enables significant saving potentials regarding CAPEX and energy use. However, the combination of two already integrated apparatuses to form the RDWC leads to a highly complex process behavior that is characterized by the interaction of phase equilibria, reaction kinetics, enzyme stability, mass transfer and component split in the dividing wall section. Employing an enzymatic catalyst allows a side product free reaction at low column temperatures. It is thus especially promising for reaction systems where conventional chemical catalysts lead to significant side reactions or require very high temperatures that would make the employment of a reactive distillation otherwise impossible.

To successfully operate such a complex apparatus, a reliable process model as well as a comprehension of the underlying process mechanism is required. The following question come to mind: What are the differences to lesser integrated setups? What do we need to know to successfully start up and operate an eRDWC? Can the dynamic and steady state behavior be predicted? How stable is an enzymatic catalyst when employed in a reactive dividing wall column?

This presentation aims to answer these questions by presenting a combination of theoretic ideas, mathematical model and experimental data. First, the basic concept of the eRTWK is presented and differences to lower integrated alternatives are discussed. A rigorous eRTWK model, based on Aspen Custom Modeler, is used to show what assumptions and equations are required to include the additional complexity generated by the higher degree of integration. Experimental data that have been measured at a pilot plant at the institute of process and plant engineering at the Technical University Hamburg, are shown to give the audience an idea of how to operate an eRDWC, what to take in mind when starting it up and what surprising mechanism of enzyme deactivation can occur in this setup. Concluding, the experimental data will be used for a model validation.

Membrane Assisted Distillation

Conference: Process Development Symposium Europe

Year: 2019

Proceeding: 2019 Process Development Symposium Europe

Group: 2019 Process Development Symposium Europe

Session: Process Intensification

Time: Tuesday, May 28, 2019 - 11:00am-11:30am

Authors:

Leva, L., Deltamem AG

Separation processes account for about half of the energy consumed in the process industries. Distillation is the most widely used separation technology and also the highest energy demanding separation unit operation. The membrane-assisted distillation (hybrid) processes have been proved to be a sustainable alternative to the conventional ternary and pressure swing distillation, traditionally used for splitting azeotropic mixtures. Alternatively, the hybrid system can increase the capacity of an existing distillation process for the same amount of energy, while still meeting the financial challenges of the retrofitting project. Pervaporation hydrophilic membranes are able to selectively remove water and methanol from organic solvents, requiring only the enthalpy corresponding to the fraction of water that permeate through the membrane. Different configurations are possible for the membrane-assisted distillation, for example the pervaporation section can be integrated into the distillation to reduce the number of trays and decrease the reflux ratio, it can be introduced in front of the distillation process to split the azeotropes before distillation or it can be installed on the top or on the bottom to achieve final product specification.

The benefits of pervaporation-based hybrid processes are known since long time, however more robust and reliable membranes, developed in the recent years, made the hybrid distillation technologies economically more attractive, because of lower membrane replacement cost. In this presentation we will introduce industrial applications for the dehydration of alcohols, ethers and the ability of the membrane to separate methanol from other organic solvents. The cases studies will show how pervaporation / vapour permeation hybrid process contributed to the process intensification strategy, by bringing significant improvement in decreasing equipment size, increasing production capacity, reducing energy consumption and enhance process plant modularization.

How Modern Digital Design Approaches can Help Realise the Potential of Process Intensification

PROCESS DEVELOPMENT SYMPOSIUM EUROPE

Conference: Process Development Symposium Europe

Year: 2019

Proceeding: 2019 Process Development Symposium Europe

Group: 2019 Process Development Symposium Europe

Session: Process Intensification

Time: Tuesday, May 28, 2019 - 12:00pm-12:30pm

Authors:

Matzopoulos, M., Process Systems Enterprise Ltd.

Process Intensification (PI), which aims to dramatically improve manufacturing processes through the application of novel process schemes and equipment, is not a new concept. PI goes beyond the incremental improvements achieved through optimising existing equipment and process schemes, by, for example, combining processing phenomena into fewer and more-integrated processing units in order to achieve step changes in energy efficiency, capital and operating costs and environmental impact. However, despite its obvious potential benefits, PI has yet to transform the process industries, partly because of the perceived risks of bringing new and unproven technologies to market in a conservative industry where mistakes can be costly.

A key challenge is that intensified processes are by definition novel and unproven, as opposed to less-efficient processes that have been well-understood for many years and therefore carry less risk. A traditional approach to process developments dictates that new processes require extensive construction of prototypes and pilots. However even exhaustive pilot testing still leaves open questions of operability and reliability, and a lack of systematic quantification of the effects of poor performance or failure, as well as the usual (e.g. scale-up) general technology risks associated with implementing new processes. There is also a perceived lack of design tools and data to develop intensified processes, and lack of a generalised workflows for dealing with the complexity of intensified, integrated modular systems.

All of this means that significant advantages can be realised from applying emerging digital design approaches that allow rapid and systematic exploration of the process decision space and rigorous quantification and management of technology risk. Digital design employs a model-based approach coupled closely with targeted experimentation. Experimentation is used to support the construction of a high-fidelity predictive model (or 'digital twin' in digital design terminology); once a model of sufficient accuracy is established, the digital twin, rather than the experimental data, is used to optimise the process design and operation.

This presentation describes, with brief illustrations, the established digital design techniques, technologies and workflows that can be applied across the intensified process development lifecycle to accelerate development and manage risk systematically. Specific topics include: capturing novel IP in high-fidelity models; validation of models against experimental and pilot data using integrated design and experimentation workflows that minimise experimentation time and cost; the application of global system analysis (GSA), a key technology for exploring the design and operational decision space, understanding sensitivity to key process parameters and quantifying and managing uncertainty and risk; and steady-state and dynamic optimisation for determining the optimal the process design taking into account operability issues or complex operating schedules.

Machine Learning Meets Production

Conference: Process Development Symposium Europe

Year: 2019

Proceeding: 2019 Process Development Symposium Europe

Group: 2019 Process Development Symposium Europe

Session: Case Studies & Lessons Learned

Time: Tuesday, May 28, 2019 - 1:30pm-2:00pm

Authors:

Büttner, N., Merantix AG

Research on machine learning topics is rapidly advancing in breadth and depth promising break-through changes on how companies operate. How to translate these cutting-edge technologies into real business value is only in its infancy. MX Labs brings cutting-edge academic research to life in applied machine learning projects across industries like healthcare, logistics, automotive and chemicals. As an example, for clients like Volkswagen, Bosch and TÜV, we have created powerful solutions for quality control in production processes, image recognition systems for autonomous driving as well as automated inventory.

This session will focus on:

- introducing examples of state-of-the-art machine learning technologies

- giving insights in how to leverage machine learning for generating real business value

- showcasing some current relevant industry applications

Status of and Perspectives for Power to Products

PROCESS DEVELOPMENT SYMPOSIUM EUROPE

Conference: Process Development Symposium Europe

Year: 2019

Proceeding: 2019 Process Development Symposium Europe

Group: 2019 Process Development Symposium Europe

Session: Case Studies & Lessons Learned

Time: Tuesday, May 28, 2019 - 2:30pm-3:00pm

Authors:

van der Made, A., Shell Global Solutions International

Increasingly, our energy system will become electrified because costs of renewable energies like solar PV and wind continue to decrease. However, whereas the traditional fossil fuels like coal, oil, and natural gas offer 24/7 availability, wind and solar energy are only available intermittently – with fluctuations varying from seconds to seasons. I will explore the use of cheap renewable electricity to create carbon-based fuels and chemicals with a zero-carbon footprint from water and CO_2. Technically such "power to products" (PTP) or "liquid sunshine" schemes are feasible, yet economically they are far from being attractive. Also, often such schemes cannot lay claim to having truly a zero-carbon footprint. I will discuss what technical breakthroughs are required to make PTP schemes being zero-carbon and economically attractive.

Using Power-to-Ammonia to Transform Fluctuating Wind Power into Base-Load

PROCESS DEVELOPMENT SYMPOSIUM EUROPE

Conference: Process Development Symposium Europe

Year: 2019

Proceeding: 2019 Process Development Symposium Europe

Group: 2019 Process Development Symposium Europe

Session: Case Studies & Lessons Learned

Time: Tuesday, May 28, 2019 - 2:00pm-2:30pm

Authors:

Hoelling, M., Hamburg University of Applied Sciences

The energy revolution in Europe is focused on the installation of renewable energies, which are mainly wind power and solar power. Due to their strong dependence on wind and sun, their production is highly fluctuating and not matching the consumption profiles of industry and households. With the on-going shut-down of base-load power plants, like e.g. coal fired power plants, the renewable energies have to be transformed into base-load by means of energy storages.

In this case study, a 3 MW wind turbine is combined with a Power-to-Ammonia process as an energy storage. In times of high electricity production, hydrogen is produced by an electrolyser. Due to the low volumetric energy density of hydrogen, it is not directly used as the energy storage. Instead, it is mixed with nitrogen to react to ammonia in a modified of Haber-Bosch-process. The reaction parameters for this exothermic reaction are chosen to be about 155 bar and 420 °C. Ammonia is separated by condensation at the reactor outlet and, since it can be stored as a liquid, it shows a huge advantage in terms of energy density (approx. 3150 kWh/m^3). In times of low electricity production, the energy storage is discharged, i.e. ammonia is fed to the same reactor but at a low pressure and a higher temperature. Thus, ammonia is split up into nitrogen and hydrogen, which can be used in a fuel cell. The heat for the reverse reaction is delivered by the combustion of a part of the hydrogen. The overall efficiency of this Power-to-Ammonia process is found to be about 30%. The results are compared to other storage technologies in terms of efficiencies, storage size, investment costs and finally electricity generation costs. For a long-term energy storage, which is capable of transforming a wind turbine into base-load, the Power-to-Ammonia process is looking very promising.

Digital Transformation in the Process Industry - Opportunities and Challenges

Conference: Process Development Symposium Europe

Year: 2019

Proceeding: 2019 Process Development Symposium Europe

Group: 2019 Process Development Symposium Europe

Session: Digitization

Time: Tuesday, May 28, 2019 - 4:00pm-4:30pm

Authors:

Hinsen, U., BASF

Some companies in the chemical industry have spent a lot of money to merge data. If these data have a sufficiently good quality, it is possible to achieve targeted savings with modern methods of data analysis. In the lecture, the prerequisites that must be fulfilled in order to achieve savings with data are entered. It introduces a "profit per hour" approach and explains how to achieve significant savings with this approach.

How to Maximize Profit per Hour in Process Industries

Conference: Process Development Symposium Europe

Year: 2019

Proceeding: 2019 Process Development Symposium Europe

Group: 2019 Process Development Symposium Europe

Session: Digitization

Time: Tuesday, May 28, 2019 - 3:30pm-4:00pm

Authors:

Froese, T., atlan-tec-Systems

Some companies in the chemical industry have spent a lot of money to merge data. If these data have a sufficiently good quality, it is possible to achieve targeted savings with modern methods of data analysis. In the lecture, the prerequisites that must be fulfilled in order to achieve savings with data are entered. It introduces a "profit per hour" approach and explains how to achieve significant savings with this approach.

AUTHOR INDEX